YEAST AND SUGAR BALLOON

Written by

Read With You Science Academy

Read With You

The characters and events portrayed in this book are fictitious. Any similarity to real persons, living or dead, is coincidental and not intended by the author.

Published by Read With You Publishing
Designed by Read With You Science Academy
Read With You and associated logos are trademarks and/or registered trademarks of Read With You L.L.C.

ISBN-13: 978-1-949391-78-7
ISBN-10: 1-949391-78-7

Second Edition April 2021
Printed in the United States of America.

Welcome to the

READ WITH YOU SCIENCE ACADEMY!

Today we will do an experiment called

YEAST AND SUGAR BALLOON

Before we begin, let's review some important words:

yeast:
A type of fungus used when baking.

gas:
A substance that is like air and has no fixed shape.

to produce:
To cause something to exist or happen.

Remember:
Be a careful scientist! Ask an adult to help you with the experiment.

SAFETY FIRST

QUESTION:

What I want to know.

What water temperature helps yeast to produce the most gas?

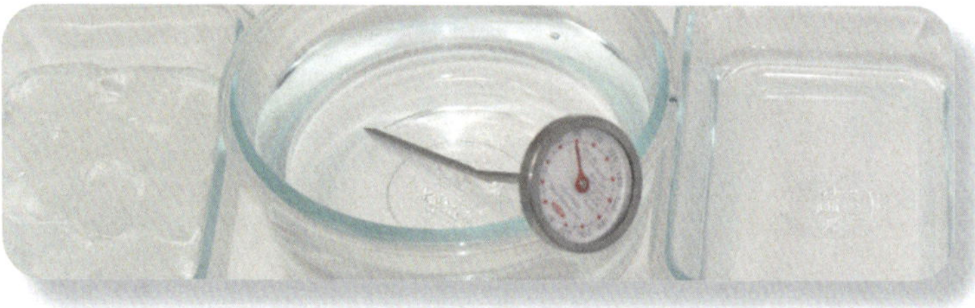

- ☐ 41°F (5°C) cold water
- ☐ 77°F (25°C) room temperature water
- ☐ 104°F (40°C) warm water
- ☐ 122°F (50°C) hot water (BE CAREFUL!)

In this experiment, you will observe the amount of gas produced by the combination of sugar, yeast, and water.

HYPOTHESIS:
What I think will happen and why.

I think **hot** water will produce the most gas because…

I think **warm** water will produce the most gas because...

I think **hot** water will produce more gas because...	I think **warm** water will produce more gas because...
Hot water has steam and produces more gas.	It is not too hot, not too cold. The temperature is perfect!
Hot water melts sugar faster.	Yeast is most active in warm water.

What do you think?
Circle your hypothesis!

EXPERIMENT:

How we will test our hypothesis.

THINGS WE NEED

- Ice cubes
- Measuring spoons
- Measuring cups
- Marker
- Funnel
- Thermometer
- Tape measure
- Gloves
- 4 empty plastic water bottles
- 8 balloons
- Towels to clean up
- 2 cups + 2 tbsp. dry **yeast** (rapid rise dry yeast works the best)
- 1 cup + 1 tbsp. of white sugar
- 2 cups each of hot, warm, room temperature, and cold water

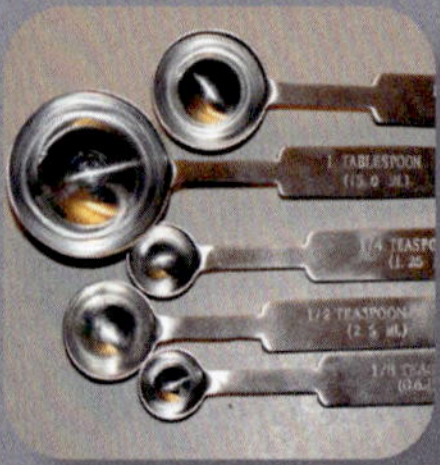
measuring spoons

funnel

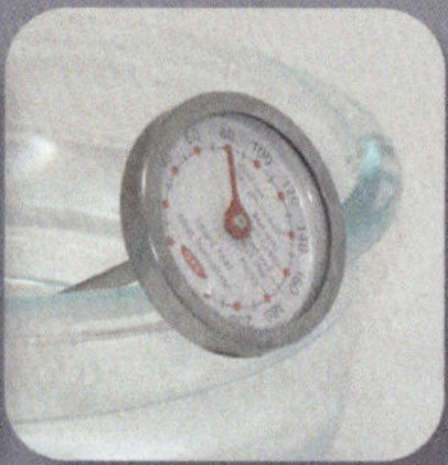
thermometer

tape measure

measuring cups

balloons

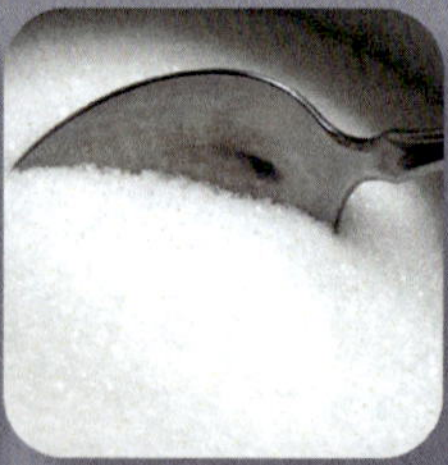
sugar

dry yeast

PROCEDURE:

The steps to complete the experiment.

Step 1

- Label each bottle with a different temperature: 41°F (5°C), 77°F (25°C), 104°F (40°C), and 122°F (50°C).
- Use the funnel to add 4 tablespoons of yeast and 2 tablespoons of sugar to each bottle.

- All 4 bottles should be labeled and contain the yeast and sugar.

Step 2

- Put on your gloves.
- Use your thermometer to measure the temperature of the hot water.
- You may need to add some room temperature water to lower the temperature to 122°F (50°C).

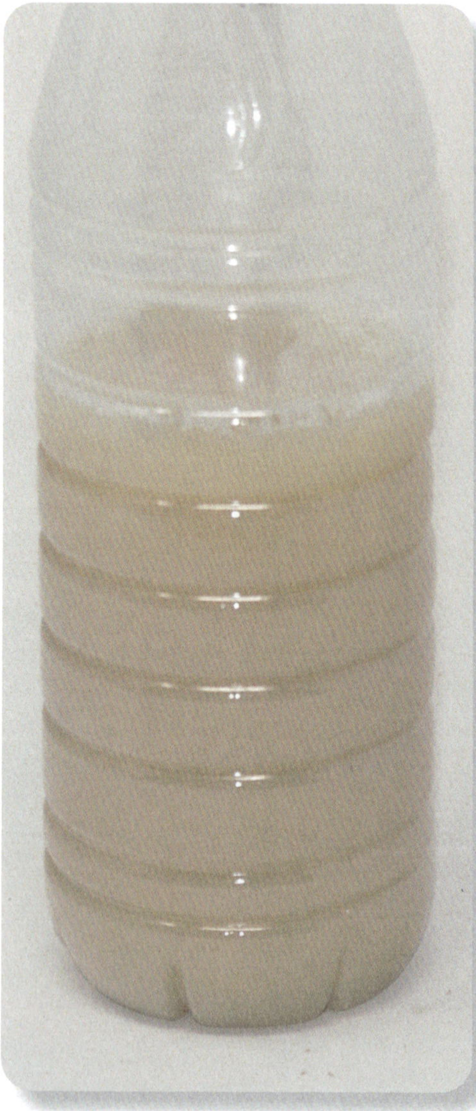

- Add 2 cups of this water to the bottle labeled 122°F (50°C) using the funnel.
- Remember: Be a careful scientist! Wear gloves when you work with hot water.

Step 3

- Quickly stretch a balloon over the top of the bottle after you add the water.
- Gently shake the water bottle.

Step 4

- Repeat Steps 2 and 3 with the other 3 water bottles.
- The water temperature should match the temperature you wrote on the bottle in Step 1.
- You can use ice cubes to cool the water to 41°F (5°C).

Step 5

- Use your tape measure to **measure** the widest part of each balloon.
- Record your measurements in your science journal.

Write down what you see

In your science journal,

write your results!

- What was your hypothesis? (Look on page 5 of this book!)
- Was your hypothesis correct?
- Which water bottle had the most bubbles? Look at the temperature you wrote on the bottle in Step 1 and write it in your journal.
- What was the size of the largest balloon?
- What was the size of the smallest balloon?

Let's do it again!

When you added water to the yeast and sugar, **gas** formed and inflated the balloons. Do you think adding more yeast and sugar will produce more gas? Let's do an experiment to find out. You will need 3 empty water bottles, more yeast, and more sugar.

QUESTION:

What I want to know.

Will adding more yeast and sugar produce more gas?

HYPOTHESIS:

What I think will happen and why.

I think adding more yeast and sugar will produce more gas because…

I don't think adding more yeast and sugar will produce more gas because...

What do you think?

Write your new hypothesis in your science journal.

PROCEDURE:
The steps to complete the experiment.

Step 1B

- Label each bottle with the number 1, 2, or 3.
- Use your funnel to add:

  4 tablespoons of yeast and 2 tablespoons of sugar to bottle #1

  6 tablespoons of yeast and 3 tablespoons of sugar to bottle #2

  8 tablespoons of yeast and 4 tablespoons of sugar to bottle #3

Step 2B

Measure 6 cups of water. Do you remember writing down the temperature of the water in the bottle with the most bubbles? That is the temperature of water you need!

- Use the funnel to add 2 cups of water to the first bottle.
- Quickly stretch a balloon over the top of the bottle after you add the water.
- Gently shake the water bottle.

Step 3B

- Repeat Step 2B with the other 2 bottles.

Write down what you see

In your science journal,

write your results!

- What was your hypothesis? (Look on page 12 of this book!)
- Was your hypothesis correct?
- How did you do it?
- Which water bottle had the most bubbles?
- What was the size of the largest balloon?
- What was the size of the smallest balloon?

Why did this work?

Yeast is a microscopic **fungus** that changes sugar into carbon dioxide. The temperature is important. Water that is too hot kills the yeast microorganisms. Water that is too cold makes the yeast inactive. Water that is warm is best because yeast is most active when it is warm and moist. The yeast and sugar did not begin to react until the warm water was added and started the fermentation process. It changed more and more sugar into carbon dioxide, which inflated the balloon.

SCIENCE

Fermentation is when a substance is "eaten" by bacteria, microorganisms, or even yeast. In our experiment, the yeast became active when we added warm water and sugar.

The bubbles you saw in each bottle were from the yeast "eating" the sugar. The waste produced was the gas **carbon dioxide**. That is why the balloon inflated—the carbon dioxide was rising from the bottle and into the balloon.

GLOSSARY

Carbon dioxide: A gas that is produced when people and animals breathe out, or when certain fuels are burned.

Fermentation: The chemical breakdown of a substance by bacteria, yeasts, or other microorganisms.

Fungus: A living thing (mushrooms, yeast, mold) that looks like a plant but has no flowers.

Gas: A substance that is like air and has no fixed shape.

Hypothesis: An educated guess; what we think will happen and why.

To measure: To discover the exact size or amount of something.

To observe: To watch something carefully.

Procedure: The steps to complete the experiment.

Produce: To cause something to exist or happen.

Question: What we want to know.

Yeast: A type of fungus used when baking.

NEW IDEAS

How could we change this experiment?

- Use a larger bottle with a lot more warm water, yeast, and sugar.
- Use balloons that are different shapes.
- Use a sugar substitute, like aspartame or sucralose.

Can you think of any ideas to improve the experiment?

- Let the fermentation process work for 15 minutes in each bottle before measuring the balloon size.
- Use the same type and shape of bottle.

JOURNAL

Made in United States
Orlando, FL
07 November 2026

90337528R00017